This book belongs to

These are...
Healthy Snacks

Written by
Cintia Roman-Garbelotto
Illustrated by
Oana Voitovici

To both of my children: Mateo, my inspiration; and Valentina, my writing partner for some of my projects.

To my husband Rudy, for his patience while I am off writing.

This is a bagel...

I like mine with

cream cheese and honey...

It's so yummy!

This is milk...

I like mine in a cup

It gives me muscle

with no hassle!

This is cereal...

In a bowl with milk,

I have mine.

It is yummy

all the time!

This is a box of raisins...

I like this snack

on certain occasions!

We squeeze some

oranges

while we have fun!

This is peanut butter...

This is toast...

With peanut butter,

honey

and banana slices,

it is priceless!

This is a fresh fruit cup...

It has watermelon, pineapple, grapes, strawberries and blueberries!

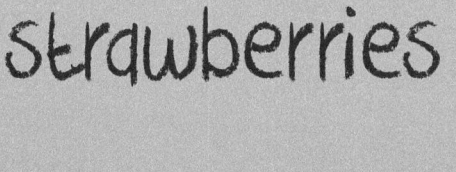

No cherries?

All these snacks

too much I like!

What about you?

What healthy snack

do YOU like?

Eat healthy!

The snacks used in this book were selected among those suggested in http://www.choosemyplate.gov as part of a healthy diet. If your child requires to follow a specific diet, please consult with your doctor and or nutritionist.